797,885 Books
are available to read at

Forgotten Books

www.ForgottenBooks.com

Forgotten Books' App
Available for mobile, tablet & eReader

ISBN 978-1-333-68270-5
PIBN 10535207

This book is a reproduction of an important historical work. Forgotten Books uses state-of-the-art technology to digitally reconstruct the work, preserving the original format whilst repairing imperfections present in the aged copy. In rare cases, an imperfection in the original, such as a blemish or missing page, may be replicated in our edition. We do, however, repair the vast majority of imperfections successfully; any imperfections that remain are intentionally left to preserve the state of such historical works.

Forgotten Books is a registered trademark of FB &c Ltd.
Copyright © 2015 FB &c Ltd.
FB &c Ltd, Dalton House, 60 Windsor Avenue, London, SW19 2RR.
Company number 08720141. Registered in England and Wales.

For support please visit www.forgottenbooks.com

1 MONTH OF FREE READING

at

www.ForgottenBooks.com

By purchasing this book you are eligible for one month membership to ForgottenBooks.com, giving you unlimited access to our entire collection of over 700,000 titles via our web site and mobile apps.

To claim your free month visit:

www.forgottenbooks.com/free535207

* Offer is valid for 45 days from date of purchase. Terms and conditions apply.

English
Français
Deutsche
Italiano
Español
Português

www.forgottenbooks.com

Mythology Photography **Fiction** Fishing Christianity **Art** Cooking Essays Buddhism Freemasonry Medicine **Biology** Music **Ancient Egypt** Evolution Carpentry Physics Dance Geology **Mathematics** Fitness Shakespeare **Folklore** Yoga Marketing **Confidence** Immortality Biographies Poetry **Psychology** Witchcraft Electronics Chemistry History **Law** Accounting **Philosophy** Anthropology Alchemy Drama Quantum Mechanics Atheism Sexual Health **Ancient History Entrepreneurship** Languages Sport Paleontology Needlework Islam **Metaphysics** Investment Archaeology Parenting Statistics Criminology **Motivational**

FIREWOODS

HEIR PRODUCTION AND FUEL VALUES

BY

A. D. WEBSTER

AUTHOR OF
SEASIDE PLANTING," "BRITISH GROWN TIMBER AND TIMBER TREES,"
"PRACTICAL FORESTRY," "THE FORESTER'S DIARY,"
ETC., ETC.

WITH TEN ILLUSTRATIONS

LONDON: T. FISHER UNWIN, LTD.
1 ADELPHI TERRACE, W.C. 2

First published 1919

ALL RIGHTS RESERVED

PREFACE

THE reasons for writing this book are:

1. That the question of firewood production and utilisation, owing to the unparalleled scarcity of coal, was never so acute as at the present time.

2. No book of a similar kind, in which the value of wood as fuel is explained, has before been written; and,

3. The author's knowledge in the matter of firewood and its utility on several of the best wooded estates in England and Scotland has caused him to relate his experience.

A. D. W.

Regent's Park,
    January 1919.

674821

CHAP		PAGE
I.	INTRODUCTION: UTILISATION OF FOREST PRODUCE	11
II.	SOURCES FROM WHICH FIREWOOD MAY BE OBTAINED	17
III.	PREPARING THE FIREWOOD	29
IV.	COMPARATIVE VALUE OF DIFFERENT FIREWOODS — PERCENTAGE OF WATER, Etc.	35
V.	HEATING PROPERTIES OF FIREWOOD—SCENTED WOOD, Etc.	39
VI.	FIREWOOD VALUE OF VARIOUS HOME-GROWN WOODS	43
VII.	FIREWOOD AND FAGGOTS—STORING, CAPACITY AND PRICE	47
VIII.	CHARCOAL WOOD—CHARCOAL BURNING—COMPARATIVE VALUE OF WOOD FOR CHARCOAL—RETURNS FROM CHARCOAL	53
IX.	CHARCOAL WOOD FOR GUNPOWDER	65
X.	FIREWOOD TABLES—WEIGHT OF FIREWOOD—MEASUREMENTS OF A CORD—PRICES OF FIREWOOD, FAGGOTS, Etc.	71
XI.	WOOD FIRES AND GRATES	75
XII.	STATE FUEL ORDER	81
	INDEX	91

LIST OF ILLUSTRATIONS

	PAGE
FAGGOT-BENCH AND CHOPPING-BLOCK	51
CHARCOAL PIT	55

A CORD OF FIREWOOD	*Frontispiece*

Facing page

THE FIREWOOD YARD	22
PREPARING FIREWOOD	32
HALF-CORD OF FIREWOOD	48
MAKING FAGGOTS	52
CHARCOAL PIT IN THE FIFTEENTH AND EIGHTEENTH CENTURIES	58
DINING-ROOM FIREPLACE (GRAVETYE MANOR)	76
FAGGOT-HOLDER (GRAVETYE MANOR)	78

FIREWOODS

CHAPTER I

INTRODUCTION

WITH the scarcity of coal, the utilisation of the forest resources of this country is now engaging the serious attention of the Coal Controller and the Timber Controller. The timber-felling going on at present all over the British Isles represents about 15,000,000 tons per annum, out of which upwards of 1,000,000 tons would be available for firewood purposes. This is exclusive of the large quantities at present in stock and the amount that could readily be procured from field and hedgerow, which, at the lowest estimate, would readily give a return of another 1,000,000 tons. Apart from this, commoners in many parts of England have the right to collect dead and dying wood, and several owners of woodlands have granted permission to residents and others on their estate to gather the fallen branches from woods and plantations—a practice that might be greatly extended in many parts of the country. Dead and dying timber will also produce a large quantity of the most useful fuel. Tree roots are also affording a considerable quantity of the best class of firewood, though the expense of obtaining such precludes any great amount being obtained. But the chief difficulty in making firewood available to the general public is transport, which at the present time, in every form, is ruinously expensive, often amounting in out-of-the-way districts to as much as the original cost of the wood. With a considerable experience of firewood and its transport on three of the largest estates in England and Wales, we have invariably found that when the distance of the timber from the consuming centre exceeds one mile, both firewood and faggots are difficult to dispose of. Trees are pruned or dressed where they fall and either dragged or otherwise conveyed to the nearest road, where they are loaded for dispatch to rail or wharf, or in a few instances to the timber yard of the purchaser. The heavier branches are dressed for firewood or other purposes and the smaller spray collected in heaps, trimmed and faggoted before being stacked for seasoning or sent direct to the distributing centres. Where the manufacture of charcoal is to be engaged in the rougher firewood need only receive a comparatively small amount of attention in the matter of pruning, and is carted direct to the position assigned for charring, which is usually a clearance in the woodland or on a piece of waste ground contiguous to the plantation boundary. But apart altogether from the felling operations, country folk might well assist in economising coal, as firewood at its present price, including delivery, is infinitely cheaper than coal. There are, however, several drawbacks to the utilising of wood as fuel, not the least, especially in towns, being the question of storage room, as a ton of timber takes up fully three times the amount of space required for the same weight of coal, while the lasting properties are considerably less, and the " trouble and bother " of stacking and preparing for consumption are by no means the least of the failings of firewood as compared with coal. There is also the question of space required for the logs indoors, and the oft renewal of a wood fire as compared with that of coal. The amount and quantity of ashes produced by a fire of wood is another drawback, especially as wood ashes, being light, are readily disseminated in the room and on the furniture. These are, however, minor details which at a time like the present will have little or no weight in the consumption of wood as fuel.

That a very considerable quantity of the produce of our woodlands, in the shape of rough trees and branches, is annually consumed for fire-lighters and fuel is not sufficiently recognised by those who are directly connected with the trade. Returns to hand from the London firewood dealers alone show that the quantity is greater than would be supposed, and the normal trade has been much increased by the exigencies of war. Vast quantities of firewood are being sent to France and Flanders in addition to charcoal and fire-lighters, with the result that there is a dearth of all these fuels at home. In many of the suburbs of London, indeed, it is impossible to purchase firewood of any kind, and much inconvenience is the result, especially as the many forms of fire-lighters are becoming rare on the market.

In ordinary times faggots and firewood are sent to the London market ready for use, the latter being bound up in bundles of the required size, and the former cut into billets ready for the fire. Large faggots, or "bavins," as they are called in Kent, have also a ready market for fire-lighting as well as being extensively used for kiln purposes.

With the present unprecedented scarcity and abnormal price of all kinds of fuel, the utilisation to the very fullest extent of all available firewood, whether of trunk, branch or root, is imperative and will alone enable us to assist in carrying out the words of the popular war song, "Keep the home fires burning."

How much of the annual fellings of home-grown timber is used as fuel is a question that is more readily asked than answered. That a very considerable quantity of the produce of our woodlands is consumed as firewood without first answering any other end will be apparent on a moment's reflection, though little is heard about it, probably owing to the fact that firewood does not appear as an article of commerce beyond the spot on which it is produced. The state of the weather will have much to do with the consumption of firewood, and during a long and cold winter the quantity sold in certain districts has been known to be fully double the amount disposed of in normal seasons. To give even a guess at the amount of firewood that is annually consumed would be hazardous, though returns from firewood merchants give a good idea of what is actually disposed of in the open market, without reference to the quantity that is sold directly from the woodlands of private estates. That the total is less than it might be is acknowledged, and it is hoped that the present outlet will be the means of fully utilising the large quantities of firewood that, for want of a market, have in the past lain and rotted in the woodlands. One great obstacle to doing this will be the comparatively large outlay in proportion to the return, but when the raw material is on the spot it seems a pity that it cannot be fully utilised, and especially at present when other fuel is scarce and expensive.

CHAPTER II

SOURCES FROM WHICH FIREWOOD MAY BE OBTAINED

THERE are four principal sources from which firewood can be obtained: (1) Where tree-felling of home-grown timber has recently been carried out; (2) from field and hedgerow trees; (3) from dead or stag-headed trees, and (4) from roots of trees that have been left in the ground. From these sources combined it is calculated that fully 2,000,000 tons of excellent firewood can be obtained. Labour is scarce and expensive, but coal is scarce too, and the principal difficulty in connection with preparing firewood for the grate will be the initial cost of so doing. This can, however, be greatly minimised by employing lame and partially disabled men to carry out the by no means hard work attending sawing and splitting the firewood into logs of convenient size for fuel. But better still, movable saw-mills might cut up the wood into blocks in the plantations where the trees are felled, the delivery to the various places of consumption either being by the purchaser's own horse or by the State at a reasonable charge per load. On several private estates with which the writer has had to do, delivery of the firewood in an unconverted state was largely carried out by the owner's teams of horses, the cost of so doing being only a few shillings per load which varied with distance and condition of the roads. Farmers, as a rule, carted their own firewood, and in some cases cottagers employed the farmer to deliver their lots at a small fixed price per load. Centres of distribution would also be a great aid in the dissemination of firewood, in such cases preference of purchase being given to dwellers in the neighbourhood.

Firewood from Field and Hedgerow.—Apart altogether from trees that are of plantation growth, the number of such as grow by field and hedgerow is very considerable, and should be largely utilised in our present unprecedented demand for all classes of timber and firewood. Everywhere one travels in the home counties hosts of big elms, oaks, and other trees may be seen occupying positions from which they can well be spared, whether the good of remaining specimens or of the surrounding land and fences be taken into account. This could be done without in the least altering the general appearance of the adjoining country, but, if carefully carried out, with the greatest advantage to the trees that may be left as standards as well as to the adjoining land. In many cases trees on fields and along hedgerows are much too close, and a thinning out would greatly facilitate an increase of timber in those that are left growing, as also in a better developed and more ornamental appearance. And what a quantity of useful timber and firewood could be got in this way, for it has been estimated carefully that in one part of Kent alone a million cubic feet of elm and oak could be spared

SOURCES FIREWOOD MAY BE OBTAINED FROM

from field and hedgerow. Take, as an instance, the fields adjoining the road leading from Bromley to Farnborough and onwards, and some excellent elm and other trees could be removed without in the least interfering with the amenity of the adjoining landscape, but with infinite benefit to the trees that would be remaining.

We have here a vast, hitherto untapped source of fuel, which, with the permission of the landowners and farmers, might yield a great quantity of the needful firewood. Timber cultivated by field and hedgerow is usually rough and branchy, so that, taken tree for tree, the yield of firewood from that grown in the open, as compared with that from woods and plantations, is fully double and of better lasting quality. Road surveyors are well aware of the damage that is being occasioned to public thoroughfares by overhanging trees, and a little co-operation with adjoining owners would be the means of procuring, at small cost and removable by good roads, thousands of cords of useful firewood. But the main quantity will be procurable from hedgerow and field, and as this is usually of rough quality owing to being grown in isolated positions where the branches got room for free growth and development, is well adapted for firewood purposes. In the removal of hedgerow and field timber on one of the largest English estates with which the writer had to do, a strict account was kept of the quantity of firewood and faggots that were obtained from each tree. The general run of trees were such as contained an average of 39 cubic feet, and were mainly composed of elm, oak, beech and alder, in the proportion of 58 per cent. of elm, 22 of oak and 20 of the remaining two. Taking a general average of nearly four hundred trees, the yield of firewood was just over three-fourths of a cord for each, all timber under 9 inches being included as firewood. The quantity of faggots worked out at twenty-seven per tree. From this it will be seen what a vast quantity of fuel can be obtained from trees that occupy positions in fields and hedgerows. The removal of these trees, if carefully and judiciously gone about, would, as before said, be, in many parts of the country at least, highly beneficial to the remaining standards as well as the adjoining agricultural land. In many parts of the country where felling operations are being carried out on a big scale, the main drawback to fully utilising the firewood is inaccessibility and cost of removal, which is just the reverse with farm and roadside trees, these being removable at the least possible expense owing to the presence of good roads.

Dead and dying trees.—Throughout every part of the country are to be found numbers of diseased, " stag-headed " and dead trees, the timber of which could readily be converted into useful firewood. Dead wood, if not actually rotten, makes excellent firewood, which, owing to its sapless condition, and being bone dry, can be used when removed from the trees. Large quantities of such are available at little expense from our park and woodland trees, and in the London area alone hundreds of tons of the most valuable firewood could be pruned from trees growing in such situations. In one park alone that is known to the writer, it is computed that five hundred loads of wood could be thus obtained; and in another, equally accessible, treble that quantity could be removed from old and stag-headed trees. Taking one instance with which the writer had to deal where the trees were old and with much dead wood at top, the average quantity of firewood procured per tree operated on was one and three-quarter loads, but this amount would be greatly exceeded in the case of existing park and garden trees. There can be no doubt that the removal of dead and dying wood from trees has a most beneficial effect on their health, and also in reducing the breeding-places of injurious insects and fungus growth. The appearance of trees from which the dead wood has been removed is distinctly improved.

Tree roots as firewood.—Firewood of great value and in considerable quantities could be procured by the unearthing and splitting up of old tree roots. Those of the oak and ash are particularly valuable in this way, though in special cases the roots of the beech, elm and Scotch pine have all been successfully utilised as fuel. The cost of labour in such cases is the main drawback, and especially at present, when wages are abnormally high and workmen fully occupied in other directions.

In not a few old plantations, however, the removal of roots from the ground that is to be replanted will be a necessity, and it is with such cases we have principally to deal. That the root forms a large proportion of every tree will be readily admitted, and in the case of the Scotch pine, ash and some other species is the most valuable for fuel, whether considered in the light of lasting or heat-giving properties. A century and more ago, when large numbers of trees were annually felled for the uses of the Navy and Mercantile Marine, tree roots, especially of the oak and pine, were largely used as fuel, and every farmer and cottager was provided with the necessary tools for splitting and cutting up the roots. In many of the peasant houses, particularly in Ireland and the north of Scotland, firewood-splitting equipment will be found, the tools including a short cross-cut saw, hand saw, heavy hammer and iron wedges. From the Irish bog-lands large quantities of excellent fuel both for fire-lighting and heating are obtained, and principally from the roots of long-submerged oak and native pine. With improved knowledge and the use of modern appliances the removal of large roots and converting these into firewood is now a comparatively easy matter as compared with the labour that attended such work some years ago. Many trees, such as the Scotch pine, beech and other hard wood species, are what is termed shallow or surface rooters, and so offer but small resistance to be pulled out, as is noticeable by the numbers of such that become uprooted during stormy weather. The simplest method is to remove the earth from around the base of the stem, sever the larger roots with a pick or stubbing axe, and, aided by the trunk, to which a rope is usually attached, pull it over and out by the aid of a horse, a

THE FIREWOOD YARD

SOURCES FIREWOOD MAY BE OBTAINED FROM 23

little under leverage with an iron-shod pole assisting greatly in the operation. From several large areas of ground in Kent and Bedfordshire the writer has had trees, roots and all removed by this method, which is simple and inexpensive, particularly if the work is carried out before the trunk is severed from the root, the leverage thus obtained simplifying the operation to a great extent. A stout rope attached to the stem of the tree at about three-fourths of its height acts as a powerful lever in uprooting. With the present unprecedented scarcity and abnormal price of fuel, nothing in the shape of firewood should be wasted or left in the ground to rot that will help to tide the nation over the crisis. Tree roots that are left in the ground are not only a harbour and breeding-place for insects and fungus pests, but they prevent the land from being successfully replanted or otherwise cultivated. Hundreds of acres of forest land are now being cleared of timber, and in the replanting of such the removal of all roots from the soil will not only be imperative but a first necessity.

Blasting tree roots.—Blasting by gunpowder or dynamite is not only the most expeditious but also the cheapest method of clearing away tree stumps and large logs. In preparing to blast a stump great care must be exercised to bore the hole in the right place and not to use too much explosive. For blasting powder the hole should be $1\frac{1}{2}$ inches in diameter, and should penetrate to the centre of the stump. It must not be too low down, lest the bottom should blow out and the force be expended in shattering the ground instead of the stump or log. In selecting the spot to bore for the powder, choose the hardest part of the root and ensure an equal thickness of wood all round, and even splitting of the log will be the result. The following is a good way of putting in the powder: For large stumps of from 2 to 4 feet in diameter about $3\frac{1}{2}$ inches depth of coarse blasting powder should be inserted in the hole $1\frac{1}{2}$ inches in diameter. The end of the fuse should be put into the centre of the powder, and left protruding for 15 inches outside the hole, which is filled with dry sand, consolidated, or packed around the fuse by means of a coarse iron wire. The outside end of the fuse should be teased out and lighted with a match, and as it will require over a minute for the fire to reach the powder, time is given for the operator to find a place of safety.

Burning tree stumps when not required as firewood.—With a 2-inch auger bore a vertical hole in the centre of the stump from the top towards the bottom. In the side of the stump, near ground-level, bore a horizontal hole towards the centre, so as to open into the vertical hole, drop some fire down the vertical hole, and if the wood is at all dry the draught of air entering by the horizontal hole will, like the draught of a chimney, maintain the combustion of the fire in the centre, until this slowly spreads and ultimately burns away the stump.

Another and equally simple method of destroying stumps of trees is as follows: In autumn bore a hole 2 inches in diameter and 18 inches deep, put in $1\frac{1}{2}$ ozs. of saltpetre, fill with water, and plug up close. In the following spring put in the same hole half a gill of kerosene oil and then light. The stump will smoulder away without blazing, down to every part of the roots.

American method of blasting.—At Studley Horticultural College, Warwickshire, the American method of blasting was successfully carried out and reported upon by **Mr. A. P. Long** as follows:—

A hole is bored with a long auger or crowbar in a sloping direction from one side of the stump to its base, generally from $2\frac{1}{2}$ feet to $3\frac{1}{2}$ feet deep. The bore-hole is cleaned out, and a number of dynamite cartridges inserted, each being firmly pressed home by a wooden rod. A primer cartridge containing a detonator is then placed on the top of these, and the bore-hole is filled with clay and tightly rammed. The primer is either connected directly with a safety fuse, or to a high-tension battery, by a cable, and is afterwards fired. As dynamite strikes downward as well as upward, the effect of the explosion is that the roots and stump are all either ejected or loosened, so that they can be easily removed by hand.

The American method is less costly and more speedy than such as have been adopted in England in the matter of removing tree stumps. If there is no man on the estate qualified to handle explosives an expert must be employed at about £1 per day, besides travelling and hotel expenses. Three men — an expert and two labourers — can bore holes and blast thirty sound stumps per day easily. If the stumps are hollow in the centre, two or three bore-holes are necessary for each stump, and in that case twenty only can be blasted during the day. Taking the pre-war wages of two labourers at 2$s.$ 6$d.$ each per day, the cost of boring and firing averages 2$\frac{1}{2}d.$ per stump, exclusive of the expert's fee. The expert's fee increases the cost to about 2$s.$ per stump.

The explosive used is Nobel's dynamite, in the form of cartridges, costing 9$\frac{1}{2}d.$ per lb. The average quantity used for each stump is between 2 lbs. and 3 lbs. (about twenty to thirty cartridges), so that the cost of the explosive is not more than 2$s.$ 6$d.$ per stump. The detonators and fuses required only cost a few pence. Summing up, the cost per stump is:

	$s.$	$d.$
Expert's fee	2	0
Cost of boring		2$\frac{1}{2}$
Cost of explosive	2	6
Detonators and fuse		9$\frac{1}{2}$
	5	6

Misfires and partial removal of stump may require fresh borings and further charges of explosive, thus increasing the cost. By employing a skilled estate hand capable of using explosives instead of an expert, the expense, however, is greatly diminished.

By the old method of grubbing and jacking, stumps were removed at Studley some time ago at the high cost of about £2 5$s.$ each butt, and even then success was only partial. In

SOURCES FIREWOOD MAY BE OBTAINED FROM 27

another case, on an estate in Norfolk, where an old pasture was converted into a plantation of mixed trees, trenching at the cost of £18 per acre had to be resorted to on account of the presence of roots and stumps of old trees. In this case it would have been much cheaper to have removed the stumps by blasting. The demonstrations at Studley showed that both sound and unsound stumps could be successfully blasted, and whole trees—an apple and an oak—were also uprooted by the same method with equal success, using only one bore-hole and about the same charge of explosives. The timber of the trees so treated, however, is very much split, so that blasting is only advisable when the timber is considered of little value.

The particular explosives used are unaffected by damp, and in consequence, the method is applicable to both wet and dry situations. Firing the charges was done at the demonstrations mostly by ladies, and a photographer was able to get sufficiently near to obtain photographs of the effect of the explosion without danger. The principal recommendations of this method, therefore, are cheapness, effectiveness and safety.

CHAPTER III

PREPARING THE FIREWOOD

FIREWOOD may be of any size, much depending on the quality and roughness of the trees from which it is cut. Usually, however, all timber under 8 inches in diameter is included as firewood. Dead or dying wood and such as has been injured by wind or lightning will also come under the category of firewood, and it not infrequently happens that in felling heavy elm trees, knotty, hollow, or such as have been attacked by insects or fungus, and in consequence rendered unfitted for commercial purposes, will be included as fuel. There are a few exceptions to the 8-inch diameter, such as in the case of rare and valuable small-growing woods, as box, holly, yew and laurel, these being sold by the ton weight, or up to 3 inches in diameter, this being specified at time of disposal. For all practical purposes, however, and as universally upheld by timber merchants and others, firewood will include all trunks and limbs under 8 inches in diameter.

Firewood is usually lotted as removed from the trees, the branches being cleared of twigs up to, say, 3 inches in diameter, after which the " spray " comes under the title of faggot wood.

The size of lot is arranged according to custom or immediate requirements, but usually in loads, half cords or full cords. Throughout England the cord is generally the adopted capacity, though in Scotland and parts of Wales, the north in particular, firewood and branches are disposed of by the cart or van load. The former method of lotting is to be recommended, as a cart load may vary greatly in bulk. Quality is usually taken into account in the lotting of firewood, and where large quantities have to be dealt with, in one and the same woodland, a fair mixture of the various kinds is to be recommended, as it would be unfair to, say, have one cord composed of poplar and another of oak. Small and large wood should, likewise, be mixed, the heating and lasting properties of the former being much less than those of the latter; and the same applies to rotten wood, or such as has become " foxy " or " dotty."

Firewood is usually lotted for sale where cut, advantage being taken of openings in the woodland by clearance roads if possible, but in order to keep down expenses it is rarely carted for cording without the bounds of the plantation where it is procured. Faggots, on the other hand, can be prepared on the ground where the branches lie, even amongst trees, thus saving the expense connected with transferring them either to division roads or outside the woodland boundary.

Tree roots make capital firewood by either blasting or wedge-splitting, the former being the most economical, especially when the roots are of large size. A charge of

PREPARING THE FIREWOOD

dynamite or blasting powder properly placed and ignited will split up the largest log into various sized pieces, these again being reduced by the mattock and wedge to the desired size. Dynamite need only be placed on the surface of the root, whereas blasting powder will require to be confined in holes bored into the thickest and strongest part of the log. Great care, as before stated, is necessary in boring the holes, which should be done with a 1½ or 2-inch auger in the thickest, strongest and most knotty part of the root and to a depth of 6 to 12 inches, according to the size of stump that is being operated on. Sometimes in the case of large spreading stumps of oak and beech it may be necessary to insert several charges of powder in order to shatter the wood so that it can afterwards be readily converted to firewood size. The quantity of powder used will greatly depend on the size and depth of boring, and great care should be exercised in ramming home the wadding so that as solid a filling to the hole may be brought about as possible.

The following method of blasting, which differs little from the above, has been successfully carried out on an English estate : " For large stumps, about 3 inches in depth of powder in a 2-inch hole will be sufficient. One end of the fuse is placed in the centre of the powder and the other left extending about 18 inches beyond the root. Dry sand or pounded clay is poured into the hole and well rammed down until the opening is quite full. A piece of half-inch iron should be thrust into the sand several times in order to make it compact.

The outer end of the fuse may then be split with a knife for, say, half an inch, and shredded for ease in lighting. About a minute will be required to reach a place of safety, and the fuse should be of sufficient length to allow of this being done. Great judgment will be required in selecting the spot where the powder is to be inserted, so that there may be an equal thickness of wood all round it, and it should not be too low in case the bottom blow out and the force be expended in shattering the ground instead of the stump."

Potentite is another valuable agent in the clearing of land of roots and reducing these to firewood size. Its downward action is a great recommendation, as it is equally powerful when left exposed as when inserted in the root, which in itself is a distinct advantage.

With dynamite there is also little trouble, all that is required being to place the charge or charges in a hollow part of the stump surface and cover with a piece of turf or loose sand, the greatest effect of this explosive being in a downward direction. We have used dynamite with good results and cheaply in clearing the stumps over a large area of ground in Scotland, the trees from which had been uprooted during a storm, and consisted of all kinds of hard-wood species over a hundred years old.

Small roots should be removed from the ground either by manual or horse labour, the latter being preferable, after the soil around the stem has been taken out. In so doing we have used a small triangular set of poles with block and pulleys

PREPARING FIREWOOD.

PREPARING THE FIREWOOD

attached, which, after the soil had been loosened around and amongst the roots, readily lifted the stump from its growing position. But removing the root by means of horse power is the cheapest and quickest method where the stumps are not too heavy, though several American devices—that are as yet but little known in this country—for the same purpose are both effective and to be recommended. No doubt with the clearance of tree stumps from large areas of cut-over woodlands some of the Canadian devices which are now offered for sale in this country will be wisely adopted. The quantity and size of stumps to be extracted will always be the best guide as to by which method they may most readily and cheaply be removed from the ground.

After being removed from the soil the shattered root portions are usually piled in lots in order that they may become dry and lighter before removal.

Cutting up the firewood, be it branch or root, is usually on private estates carried out by manual labour unless where a permanent saw-mill is established. This is most readily and speedily done by the cross-cut saw, though for small wood when in quantity the hand-saw or American one-man cross-cut is usually employed. The piece of timber is placed on a trestle which holds it firm as the cross-cutting is carried out, the log being moved along as each length is cut off. The size of grates in which the fuel is to be used will be the best guide as to what length each log should measure, but usually this varies from 6 to 9 inches for the ordinary room grate, to

c

12 to 18 inches for open fires and specially constructed burners. The diameter of log for one 9 inches in length should not exceed greatly 5 or 6 inches. When the block of firewood sawn from the log exceeds the above measurement it is usually split in two by a short, heavy axe or wood-cleaver. Small wood for fire lighting is chopped from any fast-burning timber such as logs of dry ash, plane, Scotch pine, larch, birch or other home-grown trees, tied in small bundles with tarred cord and stored in a dry, airy barn or loft till required for use.

CHAPTER IV

COMPARATIVE VALUE OF DIFFERENT FIREWOODS

THERE is considerable difference of opinion as to the respective merits of our home-grown timbers when used as firewood, though, generally speaking, sound ash, oak or beech would be placed at the head of the list. Recent experiments, however, conducted on an estate near London, point out that in heating qualities, at least, these commonly cultivated timbers are far behind several other less common kinds in their value as heat producers. When, as at present, large quantities of firewood are urgently required to supplement our meagre available supplies of coal and coke, the choice of timber to be used as fuel will be largely controlled by the woods that are most abundant, and which at present would mainly include Scotch pine, larch, beech and elm.

The age and quality of wood has, of course, much to do with its heat-producing qualities, such as is old, slow-grown, and thoroughly matured and seasoned, having greater lasting and heating properties than young, sappy timber that contains a comparatively small proportion of heartwood.

Irrespective altogether of the price or the quantities in which they can be procured, the timbers of some of our rarer trees not only burn most freely, but give out the greatest heat. Yew, when properly seasoned, approaches more nearly to coal than any other home-grown wood, both for heat-giving and lasting properties. It burns slowly, gives out a fierce heat, throws out no sparks, and is comparatively clean. Yew wood should be felled for at least two years before it is used for firewood. Laburnum wood, owing to its density and the large proportion of heartwood it contains even in a young state, approaches nearly to that of yew for fuel purposes. The use of hawthorn as firewood is proverbial, and in conjunction with apple and pear wood is greatly valued. It burns very slowly and almost without smoke, producing a great amount of heat. Hazel wood burns well, and is highly prized where it can be obtained in plenty.

Taking all in all, we are, however, inclined to place the beech and oak in the front rank as firewood producers. The timber is hard and lasting, gives out an even heat, and has the additional recommendation of being readily procured at a moderate price and easily split into logs. Oak, where it can be cut from seasoned timber, is hard to beat, though the smoke is bad for the throat. When the draught is perfect and the smoke finds its exit by the chimney there is little to complain of in oak as firewood.

Ash is a very quick burner, even when green; and elm, though a " dour " burner, is very lasting, and when thoroughly alight makes a pleasant fire. Few home-grown timbers, however, burn so brightly as winter-felled and partially seasoned

VALUE OF DIFFERENT FIREWOODS

plane; indeed, for a lively fire that of the London plane has few equals, but it is a scarce wood in England. Pine wood makes a quick fire on account of the resin it contains, but the sparking is dangerous. Scots fir, when old and resin-stained, makes a most desirable fire on a winter's night, and blazes with a glowing cheerfulness that finds a match in no other home-grown timber. Wood of this kind obtained from the Irish peat bogs is valuable and sells at a high price. Spruce and silver fir are almost valueless for firewood purposes. When used as firewood, the timber of Lawson's cypress—indeed most members of this family—gives off a delicious fragrance, and is highly valued on that account. Chestnut is not a desirable firewood; indeed, as a fire-resister it has perhaps no equal in the category of native woods. Birch burns quickly without giving much heat. Willow is to be recommended, but it must be dry and seasoned, but poplar is somewhat objectionable. The addition of a few pieces of coal to a fire of such timbers as the elm, sycamore and poplar—and, in fact, all timbers when in a green state—greatly improves their burning properties. Cedar wood burns with a pleasant fragrance, but is dangerous owing to its sparks and as giving off a volatile oil.

As before stated, some of these woods are not common, though on many private estates large quantities of yew, thorn, apple, pear, blackthorn, laurel and wild cherry are occasionally available as firewood.

The following list of timbers and their value for heating purposes has been added to from Gayer:—

(1) Possessing greatest heating power: hornbeam, beech, birch, Turkey oak, mountain pine, Robinia, old resinous Scotch pine, black pine, yew, laburnum.

(2) Possessing considerable heating power: maple, sycamore, ash, English elm, resinous larch, ordinary Scotch pine, oak, thorn.

(3) Possessing a fair heating power: Scotch elm, spruce, silver fir, sweet chestnut, Cembra pine.

(4) Possessing little heating power: Weymouth pine, lime, alder, diseased oak, aspen, poplar, willow.

As to the caloric value of seasoned oak firewood as compared with coal the proportion stands at two and a half tons to one ton, or, in other words, it requires fully two and a half times the amount of our best firewood to equal coal. The pleasure afforded by a wood fire is greater than that from coal.

Freshly-felled timber contains a large quantity of water, and for that reason is unsuitable for the best class of fuel; how great the amount will be seen from attached table:—

	Per cent.
Ash contains	28
Beech ,,	39
Birch	30
Elm ,,	44
Hornbeam ,,	18
Larch ,,	48
Oak	34
Pine	39
Poplar ,,	50
Willow ,,	26

CHAPTER V

HEATING PROPERTIES OF FIREWOOD

THE heating properties of firewood depend mainly on the amount of carbon that is contained in the woody fibre, as also the presence of oil or resin. To a large extent, therefore, the density of tissue and nature of its contents will determine the heating properties of our home-grown woods. For fire-lighting purposes, or where a quick and short blaze is required, small wood that is dry and seasoned is recommended; but when intense, long-lasting heat is of importance, heavy, close-grained wood of mature growth should be chosen. Specific gravity is not entirely dependent on the density of the particular timber, and for that reason it will not indicate the relative heating powers of different woods that are used as fuel. Take the example of oak, and that from a young, sappy tree cannot compare with the heartwood of an old and perfectly developed stem. Age, soundness, amount of moisture contained, as also the presence of oil or resin, have all much to do in determining the heating power in proportion to the specific gravity of the particular wood.

Except for fire-lighting, the use of wood as a fuel in this country may, unless in the case of private estates, be looked at in the way of a luxury, the heating properties of even the heaviest or most resinous timber being far behind that of coal, while the expense of preparation is proportionately great. Amongst our home-grown woods, when used as fuel, the difference in heating powers is considerable, and might for convenience in classification be divided into three groups, including: (1) those of greatest value; (2) of middle value; and (3) of little value. The former would include that of the yew, hornbeam, thorn, oak, laburnum, hazel, laurel, beech and resinous old pine wood. In the second group come apple, pear, ash, acacia, birch, elm, maple, evergreen oak and sycamore; while those of little value would include lime, alder, horse-chestnut, willow, spruce, poplar, larch and most of the pines. We have experimented with the timber of fully fifty of the less common kinds of coniferous trees, and found that matured *Cupressus macrocarpa, C. Torulosa*, and some of the Arborvitæ are of great value for their heat-giving properties. Such timbers as Weymouth pine, silver and Douglas fir were almost valueless for firewood purposes. For kindling purposes larch faggots are perhaps most valuable, followed by split pine wood, particularly that of the roots or old, submerged trunks of the Scotch. The Eastern or London plane produces excellent wood for fire-lighting purposes, and when split into small pieces over night has few equals for coaxing a sulky fire—a fact that is well known to the gardeners in charge of our London squares, who save every broken branch for that purpose. Except for the crackling and spitting and difficulty in handling, thorn faggots are very valuable as fire-lighters.

Elm wood, like that of several other trees, is more valuable for keeping up a slow constant fire than for igniting it. The cones of several species of pine are much sought after for fire-lighting, those of the Austrian, cluster, large-fruited, Coulter's, Stone and Weymouth, owing to their containing large quantities of resin, being especially valuable for that purpose, as also emitting a great amount of heat.

Scented Woods.—Some timbers emit a pleasant, fragrant odour when burning, and sandalwood is well known for its value in that way, being, for this reason, much used by the Chinese for burning in their temples. Amongst our cultivated trees several are both fragrant and odorous when used as firewood. The Lawson cypress, as also several other species of *Cupressus*, emit a pleasing odour when burning, as does also that of *Juniperus Virginiana*, both in the log and when used as fuel. The fragrance of this wood is remarkable, and in the case of a large tree that was felled by the writer could be detected at a distance of twenty yards. Timber of the Lebanon cedar is well known and sought after for the pleasing odour it emits when used as firewood. For smoking fish or bacon it should, however, never be employed, and at Penrhyn Castle a roomful of the latter was rendered unfit for eating owing to having been smoked with logs of cedar wood.

CHAPTER VI

FIREWOOD VALUE OF VARIOUS WOODS

For convenience of reference the following list of home-grown woods, with notes on their value as firewood, have been compiled :—

Acacia.—This timber is heavy and very durable when used as firewood. It emits a great amount of heat, but the smoke is somewhat objectionable.

Ailanthus.—Of no special firewood value, the lasting properties, owing to rapid growth and wide graining, being small.

Alder.—The heating properties of alder wood, even when old and seasoned, are only third rate. Alder chips, such as refuse from a clog-making establishment, make a lasting fire when used with a small quantity of coal.

Alder Buckthorn.—Highly valued for charcoal making.

Almond.—The wood of the almond is heavy and the concentric rings firmly packed, and for that reason burns for a long time and emits a fair amount of heat.

Ash.—The timber is well known for its rapid burning properties, even when freshly felled. For fire-lighting it is particularly valuable, and as firewood emits a great amount of heat.

Beech.—Beech firewood is one of the best, but it must not be used in a decayed state, when it is worthless either for rapid consumption or heating properties.

Birch.—Though a light wood, birch nearing maturity burns freely and emits a considerable amount of heat. Even when green it is not to be despised, and when once alight makes a pleasant fire.

Cedar of Lebanon.—Of no particular value for heating properties, but as it emits a pleasing aroma when used as firewood, is much sought after. It is most valuable when seasoned. It is inclined to send out sparks.

Cherry.—Somewhat difficult to start burning, but when once alight lasts long and gives out a fierce heat. Best when old and partially seasoned.

Chestnut (Spanish).—Third-rate as a firewood, but when fully alight makes a pleasant fire. Inclined to smoulder rather than burn brightly.

Chestnut (Horse).—Light, soft and of little value as firewood.

Elm.—Is justly recognised as a " dour " burner, but when partially seasoned and after getting fully lighted makes a pleasant, long-lasting fire.

Hawthorn.—This is one of the best firewoods, the lasting properties and amount of heat emitted placing it in the first rank. Either in a green or seasoned state it is valuable, and is far too little made use of as fuel when its properties in that particular are taken into account. It sends out sparks, so must be watched.

FIREWOOD VALUE OF VARIOUS WOODS

Hazel.—Hazel wood is well known for its value as firewood, and in consequence is much sought after by those " in the know."

Hornbeam.—This is one of our most valuable timbers when used as firewood. Being weighty, hard and close of grain, it emits a great amount of heat.

Larch.—Of no great value as firewood: indeed, it burns with difficulty, sparks and gives out little heat. Larch faggots for fire-lighting are, however, greatly prized in some parts of the country.

Lime.—The timber of the lime is light, difficult to get ablaze, and only third-rate as firewood. Inclined to smoulder.

Maple.—Maple wood, that of our native tree included, comes second in the list of useful fuels.

Mulberry wood is heavy, close grained and emits a great amount of heat when used as fuel.

Oak firewood, particularly that of mature heartwood, is hard to beat for lasting properties and the amount of heat produced.

Pear and *Apple* wood are valuable for firing, the heating properties being considerable; while they last for a comparatively long time.

Pine.—Most of the pine woods burn freely, that of the Scotch being especially valuable for fire-lighting when old and resin-stained. Unearthed roots, when chopped small and seasoned, are much sought after as kindling wood.

Plane.—Though rarely procurable in large quantities, the

wood of the plane is valuable as firewood, burning freely and making a pleasant, warm fire, whether in a green or seasoned state. For fire-lighting split plane wood, partially dry, is hard to beat.

Poplar wood is only third-rate as firewood, it being light and emitting but a small amount of heat. When used with coal it makes a pleasant fire, but alone smoulders away. Emits a light-coloured flame.

Spruce.—Not valuable as firewood, being inclined to smoulder rather than burn brightly. Dangerous for sparking.

Sycamore.—This wood has considerable heating properties and makes a pleasing fire. It should be dried quickly and under cover.

Walnut wood is too valuable to use as firewood, but it burns for a long time, though not brightly.

Willow is too light and wanting in heating power to be used as firewood. Like the poplar it is best mixed with coal when used as fuel. Inclined to smoulder and emits a whitish flame.

Yew.—Probably, as fuel, no other timber cultivated in this country produces such a fierce heat nor lasts so long as well-seasoned yew logs. The wood is remarkably close-grained and hard, and when once alight gives no trouble in the way of renewal for a long time.

CHAPTER VII

FIREWOOD AND FAGGOTS

STORING, CAPACITY AND PRICE

AFTER removal from the woods to the place where it is to be prepared for burning, firewood rarely receives the amount of attention that is necessary in order that the greatest value of the fuel may be obtained. Too often it is left exposed to all kinds of weather, with the result that it deteriorates in value and to a great extent defeats the object for which it was intended, and it should be remembered that timber that has become rotten by undue exposure to the weather makes poor firewood. When the firewood is not intended for immediate use it is highly desirable that, if a suitable building for storage is not at hand, the stack in which the wood is placed should be thatched, or in some way covered and kept dry. Stacking firewood is preferable to storing in a close building, as in the former case the stack or pile is merely roofed over and the air has the opportunity of entering freely by the sides and penetrating the entire mass. The method is simple and inexpensive, the wood being piled neatly and closely together in the form of a square and lightly thatched with reeds or faggots. Excessively wet weather, when outdoor work is at a standstill, is usually chosen in which to prepare the firewood for burning; when, therefore, a building—no matter how rough or of how temporary a nature—can be spared, it is advisable to store a quantity of the rough wood in readiness for cutting into the necessary sizes of logs.

For sale purposes the firewood in the woodlands is usually stacked and sold by the cord, though on many private estates it is disposed of by the heap or cart-load, and not infrequently delivered by the proprietors' carts at a fixed price to their customers, which in most cases are the farmers, house tenants and workmen. According to local custom the capacity of firewood varies greatly. A cord of wood is 128 cube feet— 8 feet by 4 feet by 4 feet; or 144 cube feet—12 feet by 4 feet by 3 feet; or 96 cube feet—8 feet by 4 feet by 3 feet. A cord of freshly-felled wood—128 cube feet—weighs from $1\frac{1}{2}$ to 2 tons, and will turn out 1000 billets of usual firewood size. The size of a log will greatly depend on the grate in which it is to be burned, and may vary from 3 feet in length, for especially prepared fire-places, to 9 inches for ordinary grates. If the wood is burned in an open hearth it will only be necessary to saw it into lengths of from 18 inches to, say, 30 inches, but in most cases the grate-dimensions will not admit of so large a size, and cross-cutting and clearing to the stipulated 9 inches in length by about half that in thickness will be found necessary. By far the cheapest and quickest way of cutting up firewood is by means of the circular saw, and though it is not suggested that in all cases special machinery should be installed for the

HALF-CORD OF FIREWOOD.

work, the hiring of a portable engine and saw-bench has been found both economical and satisfactory.

The price of a cord of firewood varies greatly with the district, accessibility, quality and demand, and has gone up fully 25 per cent. during the past two years. On an estate in Kent, twelve miles from London, the selling price before the war was 10s. per cord, but it is now 15s. and upwards, and even at this figure the demand is greater than the supply. About 5s. per cart-load was the usual pre-war price for rough firewood lying where felled. For ordinary purposes, however, wood as fuel can only compete with coal when the former is cheap and abundant and the latter scarce and expensive.

Large faggots for kiln and other purposes, 3 feet long and 24 inches in circumference when bound up, vary from 15s. to 20s. per 100, and small faggots, called " pimps " in the counties bordering London, which in pre-war times could be bought at 3s. 6d. per 100, now fetch 4s. 6d. and upwards. Before the war, owing to the making of faggots by pauper labour out of cheap foreign batten ends, home-made faggots for fire-lighting had decreased considerably in value.

Faggots for burning are of two kinds, large and small, the former being principally in use for kiln purposes and the latter for fire-lighting. Both kinds have quite a number of names according to the district in which they are made.

Small faggots are called " pimps " in Kent and adjoining counties, while the large are known as " bavins " or brush kiln faggots. Large faggots are made of the smaller branches or spray, the remains of charcoal and the rougher coppice or underwood, and tied into bundles similar in shape and size to a sheaf of wheat. They are about 6 feet long, and when tightly bound measure 3 feet in circumference.

House faggots.—These are made of the smaller branches, usually any that is too small for cordwood, or in the case of coppice wood, such as cannot be utilised for stakes, etc. They are made 5 feet long and 3 feet round and bound with a withe in the middle, about the size of a truss of wheat straw. They are chiefly used for fuel, fire-lighting and heating bakers' ovens.

Kiln faggots.—The small branches, called "spray" in the counties around London, cut from the underwood and all felled trees, is tied into bundles about the size of a sheaf of corn, with a band in the middle. Before the underwood is cut a man with a handbill prunes or "brushes" off all the side branches of the coppice to a height of 4 or 5 feet and ties them into bundles, which are termed brush or kiln faggots, and principally in use for brick-burning. All faggots are built up in lots of 25 or 50 as made and remain so till the following March, when they are quite dry and built in large stacks by the side of the nearest clearance road. Most of the brick kilns in which these faggots are burned are constructed as if for coal, only that the furnace is on the ground and the opening to this is of sufficient size to admit of the faggots, which are thrown into the furnace by a long-handled fork.

Faggot-making.—Large faggots are most conveniently made by driving four stakes in the ground, two opposite two, about

FIREWOOD AND FAGGOTS

3 feet by 18 inches apart. This forms a sort of framework into which the requisite quantity of branches can be packed and tightly bound by a length of tarred twine which had been laid on the ground before starting operations. The stakes can be withdrawn and taken to a convenient spot as the branches become used up. Small faggots or pimps for fire-lighting are usually made on a bench that stands about 2 feet in height and is provided with four uprights which keep the spray or small branches in position until they are tightly bound by means of tarred twine.

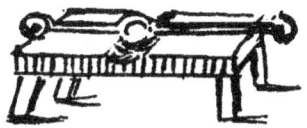 

Careful attention should be given to the lotting of firewood, as owing to loose and hollow packing there is considerable difference in the quantity of wood that is contained in various cords. When lotted by contract many unfair and sharp practices are adopted, such as placing knotty, crooked pieces in the centre, making the pile narrower at the base though of correct dimensions at the top, and crossing the wood so that the heap is partially hollow towards the centre. The same care is necessary when faggots are being made at a fixed price per hundred, as too large and branchy pieces are often packed in the centre, the contents in such cases being a few large, unpruned branches instead of a quantity of smaller, twiggy and more closely packed trimmings. It is also advisable to see that the correct number of faggots is included in the quarter, half or full hundred. Binding the faggots too loosely, whereby they are of light weight and small bulk, is another trick of the trade that must be guarded against. In districts where it is difficult to sell the faggots it is customary to prune or run the wood out as far as possible into cordwood and to burn the prunings or small spray on the ground. As greater labour is required to obtain a cord of this small wood it is found expensive to handle and comparatively difficult to dispose of profitably.

FAGGOT-MAKING.

CHAPTER VIII

CHARCOAL WOOD

SINCE the war commenced comparatively large quantities of rough and second quality firewood have been used in the manufacture of charcoal. From time immemorial charcoal burning has been carried out in our woodlands, but never before to such an extent as since called into request for the battlefields of France and Flanders. For trench heating it has no equal, and as these must be warmed without apprising the enemy of the existence of our men and so prevent soaring signals of smoke, the tent brazier is filled with glowing charcoal. On not a few private estates, throughout England in particular, charcoal is largely used for cooking purposes, and the writer for many years produced annually seven thousand bushels from the rougher and less valuable firewood that was felled on the property. For many years, previous to the war, charcoal making was a fast-dying-out industry of our woodlands, and except, perhaps, in Kent and Surrey and the English Lake district, where small quantities were annually produced for the hop kilns and iron smelting, the production of charcoal in quantity is a thing of the past.

The expert charcoal burner is now a difficult man to find,

and an independent, highly-paid workman when you have found him. Successive members of the same family in Kent and Carnarvonshire have been known to follow the occupation of charcoal burning for fully a century and a half, and it is distinctly a skilled industry, and confined to few.

Usually the men work in threes, and, having selected a piece of ground sheltered from the prevailing winds and in a position to which easy access with wood can be obtained, a rough hut is erected for the accommodation of these nocturnal workmen. Water, sand or sawdust and turf are other requisites that must be provided as the work proceeds. A couple of large tarpaulins and half a dozen straw-covered hurdles are other necessities.

From the point of economy in carting the wood to the kilns, it may seem that shifting the position of burning from one part of the woodland to another is to be recommended. Such is, however, not the case, as the hard, dry, ash-covered site, where charring has already been carried out, has its advantages, and the cost of transferring the workmen's hut and tools from one position to another must also be considered.

Several methods, largely dependent on the quantity and quality of charcoal to be obtained, are adopted, but in order to procure that of the best description the following system, which has been successfully carried out on a large estate for the past hundred years at least, is recommended. The timber carted to the charcoal yard consists of all kinds of hardwoods, preferably not under two inches in diameter.

Firewood and rough, unsaleable timber, as also inferior grades of heavy coppice wood, are mainly utilised for the production of charcoal. The wood is sawn into pieces about 2 feet long, this being the most convenient size for building the kiln, and these again split if required to some 4 inches to the side, and when a sufficient quantity for two pits has been cut up, the building of these is proceeded with. It has been found economical to burn two pits at the same time, as both can be attended to as conveniently as one, and it is unnecessary

Charcoal pit

for the men to sit up at night to watch each separately. The charcoal pits, one of which is shown in the accompanying sketch, are made of a broadly conical shape, 21 feet in diameter and about 9 feet high; the mode of construction is as follows:—

A strong stake is driven firmly into the ground and left protruding about a foot. Around this are placed small pieces of dry ash of equal length, and standing as close to the upright stake as possible; around this another layer is placed in the same manner, and this is continued until a circle 5 feet in diameter is obtained. A circle 1 foot in diameter, and having the top of the stake previously driven into the ground as centre, is next made by placing the wood horizontally on the upright pieces and side by side, the ends of each piece being placed at the circumference of the circle already made, and directed towards its centre. Layer upon layer is built in this manner until the pit is of the required height, the wood used here being dry pieces of ash 2 feet in length, but split rather smaller than the ordinary pieces. A sort of chimney is thus formed, by means of which the pit is fired. Outside the core the wood is placed on end and reclining inwards, this being continued until the pits are of the required size. When the building is complete the pits are covered with newly-cut turf, the grassy side placed innermost, beginning at the base and working towards the top, each line of turf overlapping the previous one by a few inches. The circular hole or chimney is left open for firing. Before turfing the top half of each pit it is carefully examined, and any crevices between the wood packed full of small pieces of turf and sawdust to exclude the air. The turfs are cut about 1 foot in width, and of any convenient length. The quantity required for two pits of the dimensions stated is seven loads.

When the pit is satisfactorily covered it is fired by dropping a couple of shovelfuls of burning wood and some dry pieces of pine or ash into the opening left at the top; the top turf is then put on, which effectually shuts up the chimney, and the process of charring commences. The smoke is first seen issuing from the lower half of each pit, where the chinks were not packed with sawdust, and ultimately it escapes from the whole surface.

Constant attention is required day and night during the period of burning, especially should the weather be stormy, as the wind, by striking on a particular part of the pit, causes that side to burn more rapidly, and fall in. When this occurs the hole must at once be filled in with rough logs, which had been set aside for the purpose when splitting the wood, and re-covered with turf.

When the weather is mild the pits burn uniformly, require but little attention, and produce the finest charcoal. The time required for burning will vary with the size of the pit, quality of wood, method of covering, and meteorological conditions. From six to seven days are usually required for pits of the above dimensions, but smaller kilns only covered with grass, fern and a little soil may be ready for uncovering in from two to four days. Long experience has, however, proved that by the slower process of charring the best charcoal is produced, but the cost is higher. By covering the pits with grass and fern, as is often done, a considerable saving is no doubt affected, but where turf is available, there can be no question as to its value over the former, and on the boundaries of most woodlands it is readily procurable at the cost of cutting. As the charring proceeds the turf gradually disappears until only a slight covering of burnt earth remains. When the pits have burnt out and become cool, it is found that they are reduced to rather less than half their original size.

The charcoal is extracted by means of a specially constructed rake resembling a light drag, but having much finer teeth, which, after it has become quite cold, is stored in a shed until required for use.

The very finest charcoal, superior to what is generally sold, is produced by this method. The expenses connected with making it are, however, a little heavier than usual, owing to the slower system of charring, the use of larger wood, and the extra cost of covering with turf. As to the cost of producing charcoal by the above method, this will vary greatly, much depending on the distance the wood has to be carted and on the cost of labour in the particular district.

The price paid to the charcoal burners is 7$d.$ per bushel, or about four guineas per ton, which may seem high, but when we consider that it is specialised work that is confined to a few and attended with grave risks and discomfort, the amount earned is not excessive. It should also be remembered that, previous to lighting the kilns, sufficient rough, not corded, wood has to be sawn and split and the pits carefully built and covered, not to speak of the constant attention required, both day and night, wet and dry, for from three to seven days, during charring process. The usual price for burning charcoal when the wood is corded is 35$s.$ per ton.

Fresh-felled wood is rarely converted into charcoal, the greater portion of that used being thinnings of the previous season. The proportion of wood to charcoal varies greatly, much depending on the size, quality, and maturity of timber.

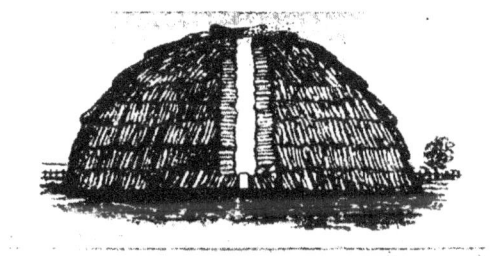

CHARCOAL MAKING IN THE FIFTEENTH AND EIGHTEENTH CENTURIES.

Having had occasion to purchase charcoal lately, I found the price, retail, to be 2*s*. 6*d*. per bushel, or in quantities of not less than a ton, £14, for that of fair quality.

From about the twelfth century onwards, Scotland, where wood was abundant, produced annually a large quantity of charcoal iron; and in 1660 the Navy Commissioners nominated John Evelyn to investigate the then denudation of forests owing to the manufacture of charcoal for iron smelting, and the following quaint extract from his report will be interesting : " Nature has thought fit to produce this wasting ore more plentifully in woodlands than any other point, and to enrich our forests to their own destruction—a deep execration of iron mills and ironmasters also." The Lorn Works, in Argyllshire, were started in 1753, and annually consumed upwards of 3000 tons of lump charcoal.

The Sussex and Kentish forests at one time supported many of the familiar charcoal burners, and right brawny and thrifty were these denizens of the woodland with their rustic, wooden huts and piles of rifted firewood, but the industry was almost a thing of the past till again called into active existence by the exigencies of the war.

Kiln-burning.—The kiln is made of brick, one course being sufficient, if bands of iron be added to strengthen the brickwork. It is usually conical in shape, 24 feet in diameter, with an equal height, and holds about forty cords of wood. The wall of the kiln is carried up nearly straight for about 6 feet, when it is gradually drawn in and made a blunt cone shape. A plate of iron is fastened on the top in the manner of a stone to an arch. Three-inch hoop-iron bands, about an eighth of an inch thick, are placed around the kiln and drawn together by means of screw-bolts and nuts. At the base and near the top are double sheet-iron doors, by which it is filled with wood or emptied of charcoal. The time required to fill, burn and empty is about three weeks. Pit-burning, for estate purposes, is, however, most commonly pursued, and has this advantage—that the charcoal can be made at any place where timber is being felled, without extra expense, save that of the cartage of the charcoal, whereas in using the kiln or retort the wood must, in most cases, be conveyed to the place where it is erected.

Comparative Value of Woods for Charcoal-making.—Amongst home-grown timber, oak, ash and beech are generally preferred for charcoal making, but the following table shows pretty correctly the proportionate relative values of the various descriptions of wood for gunpowder charcoal:

	Per cent.
Rhamnus frangula contains	27
Laburnum	25
Boxwood	24
Sweet Chestnut	23
Oak	22
Holly	20
Walnut	20
Beech	19
Sycamore	19
Elm	19
Willow	18

		Per cent.
Poplar	contains	18
Birch	,,	17
Alder	,,	17
Ash	,,	17
Hazel	,,	17
Mountain Ash	,,	17
Scotch Fir	,,	16
Larch	,,	16

Uses of Charcoal.—The uses of charcoal for estate purposes are very numerous, for horticultural, agricultural and other departments. From remote antiquity charcoal has been used as a fuel, and for many purposes it is still unsurpassed. It is by far the cleanest fuel known; it burns steadily, gives out a great amount of heat, and lasts well. On account of its smokelessness it is invaluable for cookery, and it is also admirably suited for use in greenhouse and other stoves. It is not adapted for heating apartments on account of the poisonous gas (carbonic oxide) produced in its combustion, and the danger, most apparent when the charcoal is burnt in an open *chauffer*, is not obviated by using it in a stove, as carbonic oxide has the power of diffusing through red-hot iron.

In gardening, charcoal is largely used for potting purposes, for vine borders, and for flower-beds; and in the form of dust it is the best material for packing bulbs for transmission to a distance.

Perhaps the most important of the uses to which charcoal can be put about a house or estate is that depending on its extraordinary power of absorbing gases. It is a perfect deodorant, a preservative of food and all animal substances and a valuable disinfectant. The gases most readily absorbed by charcoal are those which are most prejudicial to health and most frequently produced by putrefactive changes.

In the pores of the charcoal they are destroyed by union with the oxygen condensed from the air. The fact of its being absolutely non-poisonous and perfectly odourless puts it before all other disinfectants.

Charcoal Capacity.

The return of an acre of alder charcoal wood cut at twenty years old averages 48,000 lbs., or 600 trees at 80 lbs. each.

A cord of oak or beech wood, properly burned, yields from 40 to 45 bushels of charcoal. A rough average yield of mixed wood is 30 bushels per 128 cube feet.

Roots made into charcoal produce about 30 bushels of charcoal per cord.

Soft woods give about 53 bushels and hard woods about 37 bushels per 100 cubic feet.

A bushel of charcoal averages 17 lbs in weight.

A ton of wood will yield 40 bushels and upwards of charcoal.

The specific gravity of charcoal averages 0·20. The following notes are of interest :—

1. Heavy, close-grained hardwoods yield the best charcoal.

2. Well-prepared charcoal is black with a blue metallic tinge.

CHARCOAL WOOD

3. Brown or "foxy" coloured charcoal is imperfectly prepared.

4. When broken across the fracture should not be straight but curved, or cup-shaped.

5. It should be clean to handle, strong and firm.

6. When let fall well-burnt charcoal has a metallic or bell sound; whereas badly prepared is dull both in sound and colour.

7. Charcoal should burn without smoke, the flame being of a blue colour.

8. Slowly burned charcoal is heavier and superior in heating power to that more rapidly produced.

9. Charcoal should not be left exposed to the weather as it absorbs moisture rapidly.

CHAPTER IX

TIMBER FOR CHARCOAL GUNPOWDER

THE consumption of certain woods in the making of gunpowder has greatly increased since the commencement of the war, but when obtainable, that of the so-called dogwood—or, correctly, the black or berry-bearing alder—is preferred to any other. Owing to its comparative rarity in a wild state, this small-growing tree is usually cultivated for charcoal-making purposes. The alder buckthorn, berry-bearing or black alder (*Rhamnus frangula*), is a native shrub that is fairly plentiful in southern England, though rare in Scotland and Ireland. Confusion sometimes arises from the same popular name being applied to widely different species of plants, and this, unfortunately, is the case with the shrub in question. What is known among gunpowder manufacturers as dogwood is in reality the present shrub (*Rhamnus*), which, however, is quite distinct from the true dogwood (*Cornus*), and belongs to an entirely different family. To those who contemplate growing charcoal wood for the making of explosives, this distinction is of the utmost importance, as I have seen *Cornus sanguinea* cultivated for the making of gunpowder.

The alder buckthorn is perfectly hardy, growing freely even

in the north of Scotland, where it ripens its seed. It is usually found as an erect-growing shrub from 8 to 10 feet in height, though in suitable situations in southern England specimens fully 20 feet high, with stems 6 inches in diameter, are to be met with. The bright green leaves are oval in shape and vary, according to conditions of growth, from 2 to 3 inches in length, while the flowers are of a dull, yellowish-green and are succeeded by dark purple berries, each about the size of a pea. From a very early date the alder buckthorn has been cultivated, though not extensively, in this country for charcoal making, and the price (upwards of £15 per ton) that is paid for the wood, shows that the growing of this shrub is a profitable undertaking. At one time large quantities of the wood were produced in Sussex and other counties, the selling price being £14 per ton when peeled and tied in bundles.

The cultivation of the alder buckthorn is nearly similar to that of the osier for basket making, and the produce is dealt with and disposed of in a like manner. For soil, any good loam, inclined to be dampish, will suit it well, and an open, but not wind-swept situation should be chosen for its cultivation. The land intended for growing the alder buckthorn should be trenched the winter before planting, and a top-dressing of leaf-soil or thoroughly decomposed manure—the former preferably—will greatly assist the growth of the young plants and prevent too speedy evaporation of moisture from the soil. Young plants are not offered in quantity in our nursery catalogues, and in order to obtain a stock sufficient

TIMBER FOR CHARCOAL GUNPOWDER

to form a plantation, seed-sowing or layering old plants must be resorted to. Fortunately, by either method the plant is readily obtained in quantity, and as the seeds are produced in fair abundance and ripen freely, this method of getting up a stock is to be recommended.

The berries, after being collected in the early winter, are treated much as we treat those of the yew and holly, They are mixed with sand in order to separate the seed and fleshy covering, and the whole is sown during early spring in previously prepared beds. The seed-beds may be prepared in any shady situation out of doors, the soil being largely composed of light, sandy loam mixed with finely riddled leaf-mould. Sometimes the seeds are sown in boxes and placed in a cool frame, but we have found cultivation out of doors more satisfactory. When two years old, the seedlings should be transplanted into lines 18 inches apart and 9 inches from plant to plant. Here they may remain for another two years, after which they should be planted out permanently and headed back the following season. Rather thick final planting is to be recommended, as the shrub, being of upright growth, requires comparatively small room for development, and the best wands are produced by a close order of growth—say 5 feet from plant to plant.

Layering does not produce such upright-habited shrubs as those grown from seed, and the yield of wood per acre under exactly similar conditions of growth is greatly in favour of seedlings.

After planting, the ground should be kept free from rough-growing weeds for the first two years, the crop being cut at from six to seven years' growth, when the wands are from $1\frac{1}{2}$ inches to 2 inches diameter at butt end. Cutting and bundling is usually done by contract, but, as with the osier, it is imperative that the crop be cut over near ground level and short " stumps " without " spurs " encouraged.

As in the case of ordinary coppice wood, the buckthorn for charcoal making may be cut every sixth or seventh year, the straightest shoots, when sorted in about 5-feet lengths, being tied in bundles which are about a yard in girth. The buckthorn being a gross feeder, manuring the land after the removal of a crop has been found advantageous.

Although largely imported from Holland and other parts of the Continent, home-grown wood is preferred, as it produces a much superior charcoal for the manufacture of explosives. Unfortunately, however, home supplies are so limited that foreign wood is imported in considerable quantity, and as there was a scarcity before the war, the probabilities are that, with our greatly increased consumption of charcoal explosives, a dearth of suitable wood is now being felt.

With the present small remuneration attaching to the cultivation of coppice or underwood, which under ordinary circumstances does not exceed £4 per acre, the cultivation of the alder buckthorn is to be recommended, particularly as the price is at least quadrupled. The cultivation of this is quite simple, while the quality of soil required need not be better than that which

TIMBER FOR CHARCOAL GUNPOWDER 69

produces a crop of chestnut or hazel. In cultivating the alder buckthorn for charcoal purposes the following rules should be observed :—

1. It will not succeed satisfactorily in sandy, poor, or water-logged soils; rich, well-manured loam being preferred.

2. The ground should either be trenched or ploughed and cleared of all rough-growing weeds the winter before planting.

3. Plant seedlings or layers in the spring in lines about 5 feet apart, and the same distance from plant to plant.

4. An annual clearance of weeds and loosening of the soil between the rows of plants is recommended where a heavy crop is expected.

5. Induce the growth of stout, clean shoots by liberal feeding and clean cultivation.

6. Cut the shoots close to the ground so as to prevent the formation of long spurs and minimise the number of offshoots. Clean cutting with a sharp tool is imperative.

7. After the removal of a crop, stirring and enriching the soil is to be recommended.

8. Though found mixed with under-shrubs and in the shade of trees when in a wild state, yet the greatest quantity of the most valuable wood for charcoal making is produced in open situations.

Heating Properties of various Charcoals.—With a view to finding out which of the charcoals produced from our home-grown timbers give out the greatest heat, the following list has been drawn up from carefully-conducted experiments made on the Dalkeith property :—

Kind of wood.	Time taken to boil water.
Ash	16 minutes
Bay Laurel	17 "
Beech	11
Boxwood	18
Cedar of Lebanon	22
Elm (English)	13
Elm (Scotch and Wych)	20
Hazel	15
Holly	13
Hornbeam	14
Laburnum	27
Privet	14
Rosewood	18
Spanish Chestnut	16
Sycamore	13
Thorn	15
Yew	20

CHAPTER X

FIREWOOD TABLES

TABLE showing the weight in pounds per cubic foot of our commonly cultivated timbers in a round or unconverted state, and partially seasoned with the bark attached:—

Ash	50 years old,	73	lbs. per cubic foot.
Beech . . .	70 ,,	84	,, ,, ,,
Birch . . .	70 ,,	82¾	,, ,, ,,
Elm	70 ,,	76	,, ,, ,,
Horse Chestnut .	50 ,,	70	,, ,, ,,
Sycamore . .	50 ,,	88	,, ,, ,,

AVERAGE WEIGHT OF FIREWOOD

Cord of fresh felled firewood weighs 31 cwts.

Cord of seasoned firewood weighs 21 cwts.

This refers to mixed firewood, with little or no pine wood.

TABLE OF STANDARD MEASUREMENTS OF A CORD OF WOOD AS FIXED BY LOCAL CUSTOM

Length.	Height.	Width.	Total cubic contents.
12 ft.	3 ft.	3 ft. 6 ins.	126 cubic ft.
12 ft.	3 ft.	4 ft.	144 ,,
12 ft.	3 ft.	3 ft. 3 ins.	117 ,,
14 ft.	2 ft. 9 ins.	3 ft. 3 ins.	125 ,,
14 ft.	3 ft.	3 ft.	126
8 ft.	4 ins.	4 ft.	128

Around London the first and last measurements are usually adopted.

Table of Standard Sizes of Faggots

	Length.	Girth of centre when tied.
Large faggots	4 ft.	3 ft.
Small	9 ins.	13 ins.

Prices of Producing Firewood

Cutting out and stacking firewood per cord, 7s.
Cutting out and preparing faggots per 100, 4s. 6d.

Prices of Firewood and Faggots

Firewood, per cord	20s. and upwards.
Faggots (large)	18s. to 25s. per 100.
Faggots (small)	4s. 6d. to 5s. ,, ,,

(Prices vary greatly with district and demand.)

Specific Gravity of Various Firewoods

The specific gravity of timber varies greatly with the drying process, and, as might be expected, some of the rapid growing trees, such as the poplar and several species of pine, lose proportionately a greater amount in weight than the heavier-wooded species, as the oak and acacia. The greatest specific gravity of various timbers when freshly felled is that of hornbeam, followed by oak, Austrian pine, beech, elm, poplar, birch, sycamore, ash, acacia, willow, alder, Weymouth and Scotch pines, larch, lime, plane and spruce.

In the following table the specific gravity of the various firewoods was taken four weeks after the timber was felled and when, owing to the weather, it was in a comparatively dry state.

Kind of wood.	Specific gravity.
Oak	90
Hornbeam	90
Birch	82
Ash	81
Beech	80
Acacia	80
Sycamore	79
Austrian Pine	73
Poplar	71
Larch	70
Willow	69
Alder	68
Scotch Pine	66
Elm	62
Lime	59
Plane	58

The following short notes in connection with the treatment of firewood should be observed:—

1. Firewood should contain all the solid materials possible.

2. Should be winter felled.

3. Should be dried rapidly, the best season being that of alternate drought and heat.

4. Should be stacked and protected from wet after being seasoned.

5. Rotten wood is practically valueless as firewood. It smoulders and produces little heat.

6. Dead branches are valuable as firewood, being in every respect similar to thoroughly seasoned wood.

7. Do not cut the logs of wood too small a size, 9 inches long by about 5 inches thick being the most suitable dimensions for the ordinary grate.

8. Poking at or interfering with a wood fire is objectionable from a lasting point of view.

From private estates much useful firewood might be obtained by the felling of such trees as are absolutely unfit for timber, including rough, branchy specimens and such as are unhealthy, dead or dying. Labour being scarce on most estates, these trees might either be offered to the Timber Controller or disposed of to some firewood merchant who has regular gangs of efficient tree-fellers, and who would quickly, by arrangement, convert them into the needful fuel. The cordwood having been secured, most purchasers will find ways and means of cutting into logs for fuel.

CHAPTER XI

WOOD FIRES AND GRATES

"THERE is no fire so beautiful as a wood fire on the hearth. It is economical, too, if only in not having to remove a mass of coal ash every morning." So writes Mr. W. Robinson, of Gravetye Manor, in his most useful and sumptuous book, *My Wood Fires and Their Story*. With plenty of wood on his estate, the author thought it wrong that the best of all fuels could not be well used in the house, and so set himself to consider the problem in all its bearings. Even in London, as will be gathered from the following letter to Mr. Robinson, wood fires with properly arranged grates have been found a decided success.

"When I had rebuilt the old seventeenth-century house, which stood on this site—here in Westminster—I decided to retain the old method of heating which had evidently been in use in the old house for a number of years after it was built, namely, the old fires of wood, burning right on the hearth, with fire-dogs; being anxious to preserve the old style of hearth, and fortunately able, through your advice and with the aid of a good architect, Mr. W. F. Troup, to accomplish what I wanted.

"I had seen enough of your splendid cleanly wood fires, scenting the whole atmosphere of the house, at Gravetye, appreciating the delight and advantages derived from them, to make me want to eliminate the use as far as possible of any form of coal in open grates.

"Needless to say, the fires are the delight of every one who comes here. They are very satisfactory, and far more so than any coal fires, and give ample heat, and they are practically no trouble. The ash is seldom removed—not more than two or three times a year, with a fire in constant use. This accumulation of ash forms a bed which during the winter months is always warm, and the fire lights promptly and easily with the aid of a small faggot or pimp.

"The convenience of this form of fire in a bedroom in times of sickness is precious, for it will keep alight all through a night without any attention, whereas any coal fire has to be replenished constantly, and relit daily. On more than one occasion the same fire has been kept going thirty consecutive days and nights, not allowing it to go out at all!

"The only difficulty one suffers from in London, where log fires are used, is the small storage accommodation for the wood in houses, and following from that, some difficulty in having a sufficiency of regular supplies; the fact that hardly any one has log fires in our larger towns and cities results in

DINING-ROOM FIREPLACE.
Gravetye Manor.

there being no demand for wood and no stocks at hand. It is therefore necessary for any one with log wood fires to be able to store an ample quantity, and this, generally speaking, cannot be done. When having this house built I devoted all the space possible to the storage of logs, but it is not sufficient to ensure a constant supply. Logs are bulky. I reckon that in a space which will accommodate four to five tons of coal not more than two tons of wood can be stored."

From this and other experience there can be little doubt that wood fires are to be recommended, and though properly arranged grates for burning wood are to be preferred, yet, in an ordinary fireplace, by cutting the logs to the required dimensions, a pleasant and profitable way of heating is in the reach of all. Preferable would be the old-fashioned hearth on the floor level, but as such is rarely found in modern houses, though common enough in the Highlands of Scotland and in Ireland, the best must be made of the up-to-date grate which, if properly stoked, will be found a fairly good substitute. The great drawback is that in the ordinary grate only small logs can be consumed, whereas in the flat, open hearth tree trunks, up to 4 feet in length, may be readily burnt, with the advantage that the expenses of sawing and splitting are greatly minimised and less attention to stoking required. Another advantage is that refuse wood of all kinds, even that of a bulky nature, can be used as firing, while the consumption is much less in the case of a hearth fire than with a grate, when this is elevated several feet above floor level. Daily rekindling is also not a necessity, as by keeping ashes over the ignited wood it remains in a glowing condition until again uncovered, when it blazes quickly by the addition of some smaller pieces of firewood. But even in the ordinary grate firewood burns freely, only the logs must be cut into the required size. On the Continent much of the cooking is done by a wood or charcoal fire, where annually large supplies of both are provided for this purpose. That a wood fire is much more pleasant than one of coal is appreciated by all; while the delicious aroma given off by some of our native timbers when used in the house is quickly noticed by the visitor. There is no necessity, however, to be provided with a large, expensive fireplace, as amongst modern grates several, including the "Drawwell," are quite suited for the ordinary sized house in which firewood is used as fuel. In this particular type, which has been much used in London of late, owing to the bottom bars being only a few inches above floor level and the draught regulated at will, the consumption of blocks of wood is both pleasant and easy. Where firewood is largely made use of for fuel it becomes a necessity to provide a suitable wood and faggot shed in which the work of preparation of the firing and binding the fire-lighting faggots may be carried out, work that can often be done in wet weather when outdoor operations are at a standstill. On an estate known to the writer a shed of this kind was simply and cheaply erected of wood and

FAGGOT-HOLDER (OLD DUTCH, COPPER).
Gravetye Manor.

galvanised iron roofing. Plenty of ventilation is necessary, and the shed should be sufficiently large not only to store a goodly quantity of firewood but provide sufficient room for sawing up and splitting the logs. Faggots for fire-lighting were also made and stored in the same building.[1]

I am indebted to Mr. Robertson for the use of the accompanying illustrations.

[1] Although so intimately associated with our Christmas festival the Yule log is of Pagan origin. It belonged originally to the Saxon feast of Jul—pronounced Yule—and only passed into Christian observance at a later period, when our Barons made their hearths big enough to accommodate the entire butt end of a tree.

CHAPTER XII

THE FUEL WOOD ORDER

THE following Order dealing with the sale and distribution of firewood came into effect on the 1st of October, from which date all wood for fuel purposes is to be held at the disposal of the Board of Trade, the regulation of quantities and prices being delegated to District Divisional Officers.

Briefly the Order is that only in districts where it is cut will the wood be available, and that local fuel and lighting committees will fix the prices, which will range from 40s. to 50s. per ton. Local authorities will also have preference in securing and disposing of the firewood, which will include all timber that is unsuitable for pit-props or sawn lumber wood, from two inches in diameter and upwards. No one will be entitled to buy from a retailer more than two tons of fuel wood without a permit from the Local Fuel Overseer. In districts where fuel wood is plentiful consumers will be required to take a portion of their allowance in wood in lieu of coals, two tons of wood being reckoned as equivalent to one ton of coal.

The Order does not prevent owners of woodlands from making reasonable use of firewood for their own estate purposes, nor employees or others from gleaning in the woods and plantations. There is no fixed rate for firewood prices in the wood, but 15s. per stack for soft wood and 20s. per stack for hard wood has been mentioned.

THE FUEL WOOD ORDER, 1918

Board of Trade,
7, Whitehall Gardens,
London, S.W.1.

THE FUEL WOOD ORDER, 1918, DATED 27TH DAY OF SEPTEMBER, 1918, MADE BY THE BOARD OF TRADE UNDER REGULATIONS 2F/2G/2J and 2JJ OF THE DEFENCE OF THE REALM REGULATIONS.

THE Board of Trade, deeming it expedient to exercise the powers conferred upon them by the Defence of the Realm Regulations for the purpose of maintaining the supply of wood suitable to be used for fuel, hereby order as follows:

1. In this Order the expression "Fuel wood" means the waste lop and top of felled timber exceeding 2 inches in diameter, and any other timber unsuitable for conversion into sawn lumber or pit-wood, and waste produced in the conversion of timber at a saw-mill or factory.

The expressions "Local Authority," "Local Fuel and Lighting Committee" and "Local Fuel Overseer" have the same meanings as in the Household Fuel and Lighting Order, 1918, and the Household Fuel and Lighting (Scotland) Order, 1918.

2. Where standing timber is or has been felled the person responsible for the felling shall, except as provided in para-

THE FUEL WOOD ORDER

graph 23 hereof or unless otherwise permitted or directed in writing by the Controller of Timber Supplies, cause all fuel wood as and when produced by such operations to be collected into stacks at roadside or at some place or places convenient for removal by mechanical or other transport.

A stack shall measure 16 feet by 4 feet by 2 feet, or 8 feet by 4 feet by 4 feet, and be closely packed.

3. Subject to the provisions of paragraph 23 hereof, every person who fells timber and every person converting such timber at a saw-mill or factory shall hold all fuel wood produced at the disposal of the Board of Trade, and shall give notice thereof to the Divisional Officer of the Coal Mines Department of the Board of Trade for the district in which it is lying, and shall, if required, deliver it at roadside or other place of removal mentioned in paragraph 2, or at the saw-mill or factory, as the case may be, to the order of such Officer. No person shall offer fuel wood for sale or otherwise dispose of it before offering it to such Divisional Officer, who shall accept or refuse in writing any fuel wood or part thereof so offered to him within 28 days of the receipt of the offer. The titles and addresses of the Officers and particulars of their districts appear in the Schedule hereto.

4. No person shall sell fuel wood by retail without a licence from a Local Authority, except as provided in paragraph 23 hereof.

5. A licence to sell fuel wood by retail shall be in a form approved by the Board of Trade, and may define an area

within the district of the Local Authority issuing the licence within which the holder thereof may sell fuel wood, and may, with the consent of any other Local Authority, authorise the holder to sell within any part thereof as may be defined.

6. A licensed retailer shall not sell more than 2 tons of fuel wood to any person for consumption on any premises to which the Household Fuel and Lighting Order, 1918, or the Household Fuel and Lighting (Scotland) Order, 1918, apply during any period of twelve calendar months, except with the consent in writing of the Local Fuel Overseer of the district within which such premises are situate. The said amount may be increased or decreased by notice issued by the Local Authority on behalf of and with the consent of the Controller of Coal Mines.

7. Where a person has offered fuel wood to a Divisional Officer under paragraph 3 hereof and it has been refused, he shall not sell or dispose of such wood, except as permitted by paragraph 23 hereof, to any person other than a licensed retailer unless he himself is a licensed retailer.

8. No person shall buy or acquire or attempt to buy or acquire for consumption more than the amount of fuel wood fixed by or in accordance with paragraph 6 hereof without the consent in writing of the Local Fuel Overseer.

9. A Local Fuel Overseer may, with the consent of the Controller of Coal Mines, require consumers within his district to take fuel wood in part satisfaction of the allowance of fuel granted to them under the provisions of the Household Fuel and Lighting Order, 1918, or the Household Fuel and Lighting

THE FUEL WOOD ORDER 85

(Scotland) Order, 1918, provided that he shall not require any consumer to take more than one-third part of his allowance in fuel wood, and for this purpose 2 tons or such larger amount as a Local Fuel Overseer, with the consent of the Controller of Coal Mines, may determine shall be deemed the equivalent of a ton of coal.

10. Any additional allowance of fuel granted under Clause 11 of the Household Fuel and Lighting Order, 1918, or under Clause 10 of the Household Fuel and Lighting (Scotland) Order, 1918, or any part thereof, may be made in fuel wood.

11. Where a consumer is required to take fuel wood in satisfaction of any part of his allowance or of an additional allowance of fuel, the fuel wood so taken shall not affect the amount with which he is entitled to be supplied or which he is entitled to acquire under paragraphs 6 and 8 of this Order.

12. Where a Local Fuel Overseer has required a consumer to take fuel wood in satisfaction or part satisfaction of his fuel allowance a licensed retailer shall supply such a consumer with the amount which he is so required to take in priority to all other consumers not so required as aforesaid.

13. A Local Fuel Overseer may grant to any consumer within his district a certificate entitling that person to obtain an amount of fuel wood to be specified in such certificate in priority to all persons other than those mentioned in paragraphs 9, 10 and 12 hereof, and a licensed retailer shall supply such person accordingly.

14. The price at which fuel wood may be bought or sold

at roadside or other place of removal referred to in paragraph 2 or at a portable or forest mill shall not exceed 15$s.$ per stack for soft wood and 20$s.$ per stack for hard wood, provided that the Controller of Timber Supplies may, by notice under his hand, vary such maximum prices from time to time, either generally or for any particular district, or for any particular class or description of wood. Whether any fuel wood is soft wood or hard wood shall be determined according to the usual custom of the trade, and in the case of dispute by the Controller of Timber Supplies.

15. A Local Authority which deals in fuel wood or which has granted licences to deal in fuel wood shall fix, and give public notice of the maximum prices at which fuel wood or any description thereof may be bought or sold by retail within their district, but shall not have power to vary the maximum prices fixed under or in accordance with the last preceding paragraph. Prices so fixed shall allow for the cost of transport and delivery and for the sawing into short lengths to the extent to which such services are performed.

16. Subject to Clause 14 hereof no person shall buy or sell fuel by retail at a price exceeding the maximum price fixed in accordance with the preceding paragraph of this Order. Nor shall any person impose or attempt to impose any conditions on a sale or proposed sale of fuel wood under this Order.

17. A licensed retailer shall exhibit and keep exhibited in a conspicuous position at every place where he sells fuel wood a notice of the maximum prices in force for the time being

or of any less prices at which he is willing to sell. A hawker licensed as a retailer shall exhibit such notice on the vehicle from which he sells fuel wood.

18. The Local Authority shall determine the method, whether by measure or weight, by which fuel wood shall be sold by retail within their district and no person shall sell fuel wood by any method other than that so fixed.

19. Where fuel wood is delivered to the order of a Divisioual Officer under the provisions of paragraph 3 the person engaged in felling the timber or the person engaged in converting such timber at a saw-mill or factory shall, if he so desires, be entitled to hand the fuel wood and deliver it at such place as such Officer may direct, at a price to be agreed with such Officer.

20. A Local Authority may, on behalf of the Board of Trade, and is hereby authorised to, require all persons engaged in the sale of or dealing in fuel wood to make such returns and to supply such particulars relating to their business as the Local Authority may for the purpose of the Order require. A licensed retailer shall in particular keep a record of all sales of fuel wood made by him exceeding in value one shilling or such other value as may from time to time be determined by the Local Authority, showing the customers' names and the dates and amounts of such sales, and such record shall at all times be open to inspection by the Local Authority or by any person authorised in that behalf by the Board of Trade.

21. The Local Authority shall in all matters relating to this

Order act through the Local Fuel and Lighting Committee and Local Fuel Overseer.

22. The licensed retailers of fuel wood in any district may nominate or elect a member to be added to the Local Fuel and Lighting Committee of the Local Authority for such district, and Clause 21 of the Household Fuel and Lighting Order, 1918, and Clause 30 of the Household Fuel and Lighting (Scotland) Order, 1918, are hereby amended accordingly.

23. Nothing in this Order applies to fuel wood :—

(*a*) Sold on the ground by persons engaged in felling operations or in operating a portable or forest mill in quantities not exceeding 2 tons to any one consumer in any period of twelve calendar months at a price not exceeding the maximum fixed under paragraph 14;

(*b*) Felled by the owner on his own land for his own use or for the use of his tenants or servants, unless it is the waste of timber felled for sale;

(*c*) Collected or gleaned by persons for their own consumption in accordance with any custom or with the consent, express or implied, of the owner or occupier of the land;

(*d*) Given to workmen according to the custom of the trade or as part of their wages;

(*e*) Bought or sold for chemical purposes; provided that any person to whom wood is sold under Clause (*a*) hereof shall bring it into account in calculating the amount which he is entitled to acquire under paragraphs 6 and 8 of this Order.

24. The Board of Trade may suspend the operation of this

Order within the district of any Local Authority for such times and subject to such conditions (if any) as they may think fit. Notice of suspension shall be given by the Local Authority.

25. All Contracts other than contracts with a Government Department or with a Naval or Military Authority for the purchase and sale of fuel wood existing at the date when this Order comes into operation are hereby abrogated.

26. Infringements of [this Order are summary offences against the Defence of the Realm Regulations.

27. This Order comes into operation on the 1st October, 1918, and applies to Great Britian only.

28. This Order may be cited as The Fuel Wood Order, 1918.

A. H. STANLEY.

INDEX

A

Acacia, value as fuel, 43
Ailanthus, value as fuel, 43
Alder buckthorn, rules for cultivating, 69
    for charcoal making, 43
Alder wood as fuel, 43
Almond wood for fuel, 43
American method of blasting, 25
Apple wood as fuel, 45
Ash wood as fuel, 43
Aspen wood as fuel, 38

B

Bavins, 49
Beech, value of wood as fuel, 44
Best wood for fuel, 36
Birch wood as fuel, 44
Black pine wood as fuel, 38
Blasting, American method, 25
    tree roots, 23
    tree stumps, cost of, 26
Burning charcoal, 53
    , grates for, 75–9
    tree stumps, 24

C

Capacity of firewood, 47–52
Cedar of Lebanon as fuel, 44
Charcoal and its manufacture, 55
    burning, 53
    , commercial aspect of, 59
    , heating properties of, 69

Charcoal, notes on, 62
    , returns from, 62
    , specific gravity of, 62
    , uses of, 61
    wood, 53–63
    wood for gunpowder, 65–9
Cherry, value as fuel, 44
Chestnut wood, horse and Spanish, for fuel, 44
Coals added to a fire, 37
Commercial aspect of firewood and faggots, 47
Comparative value of wood for charcoal, 60
    different firewoods, 35–38
Cord, measurements of, 71
Converting, cost of, 72
Cost of blasting tree stumps, 26
Cultivation, alder buckthorn, rules for, 69
Cypress, Lawson's, as firewood, 37

D

Dead and dying trees as firewood, 20
Different firewood, percentage of water, 38
Douglas fir, 40

E

Eastern plane as firewood, 37
Elm as firewood, 38

INDEX

F

Faggot making, 50
Faggots, house and kiln, 50
  , price of, 47–52, 72
  , storing of, 47–52
Field and hedgerow, firewood from, 18
Firewood capacity, 47–52
  , comparative value of, 35
  from field and hedgerow, 18
  , heating properties of, 39–42
  , notes on treatment of, 73
  , price of, 47–52
  , sources from which obtained, 17–27
  , storing of, 47–52
  supplies, State and, 81
  tables, 71–4
  value of various woods, 43–6
Firewood, value of—
  Acacia, 43
  Ailanthus, 43
  Alder, 43
  Almond, 43
  Apple, 45
  Ash, 43
  Aspen, 38
  Beech, 44
  Birch, 44
  Black pine, 38
  Cedar of Lebanon, 44
  Cembra pine, 38
  Cherry, 44
  Chestnut, horse, 44
  Chestnut, sweet, 44
  Douglas fir, 40
  English elm, 38
  Hawthorn, 44
  Hazel, 45
  Hornbeam, 45
  Laburnum, 36
  Larch, 45
  Laurel, 37

Firewood, value of (*continued*)—
  Lime, 45
  Maple, 45
  Mountain pine, 38
  Mulberry, 45
  Oak, 45
  Pear, 45
  Pine, 45
  Plane, 45
  Poplar, 46
  Robinia, 38
  Scotch elm, 38
  Scotch pine, 38
  Scots fir, 37
  Silver fir, 38
  Spruce, 46
  Sycamore, 48
  Turkey oak, 38
  Walnut, 46
  Weymouth pine, 38
  Wild cherry, 37
  Willow, 46
  Yew, 46
  home-grown woods, 43–6
Firewood, weight per cubic foot, 71
Forest produce, utilisation of, 11
Fuel, best wood for, 36
  Order, State, 81–9

G

Gayer's *Heating Properties of Timbers*, 38
Grates for burning, 75–9
Gunpowder, charcoal wood for, 65–70

H

Hawthorn timber as firewood, 44
Hazel timber as firewood, 45
Heating Properties of Timbers, Gayer's, 38
Heating properties of uncommon timbers, 40

INDEX

Heating properties of various charcoals, 69
    various woods, 39–41
Hedgerow, firewood from, 18
Hornbeam timber as firewood, 45
House faggots, 50

I

Irish peat bog, timber from, 37

K

Kiln burning, 59
    faggots, 50

L

Laburnum wood, value as fuel, 36
Larch wood, value as fuel, 45
Lawson's cypress, 37
Lime wood as fuel, 45
London plane, value as fuel, 37
Lotting of firewood, 51

M

Manufacture of charcoal, 55
Maple, 45
Measurements of a cord, 71
Mountain pine, value as fuel, 38
Mulberry wood, value as fuel, 45

N

Notes on charcoal, 62
    on treatment of firewood, 73

O

Oak, Turkey, value as fuel, 38
    as fuel, 45
Order, State Fuel, 81–9

P

Pear wood as fuel, 45
Percentage of water in timber, 38
"Pimps," 49
Pine, black, as fuel, 38
    , mountain, as fuel, 38
    wood as fuel, 45
Plane wood as fuel, 45
Poplar wood as fuel, 46
Preparing the firewood, 29–34
Price of firewood and faggots, 47–52, 72
Produce, forest, utilisation of, 11
Producing firewood, prices of, 72
Properties, heating, of various woods, 39–41

R

Returns from charcoal, 58
Rhamnus frangula for charcoal, 60
Robinia, value as firewood, 38
Roots for firewood, 21
    , blasting, 23
Rules for cultivating the alder buckthorn, 69

S

Scented wood, 41
Scots fir as firewood, 37
    pine as firewood, 38
Short notes on treatment of firewood, 73
Silver fir as firewood, 38
Smoke from oak wood when used as fuel, 36
Soft woods, produce for charcoal, 62
Sources from which firewood may be obtained, 17–27
Specific gravity of charcoal, 62
    various woods, 73
Spruce as fuel, 46

INDEX

Standard measurement of a cord of wood, 71
   sizes of faggots, 72
State and firewood supplies, 81
State Fuel Order, 81–9
Storing of wood and faggots, 47–52
Stumps, cost of blasting, 26
   , tree, burning, 24
Sweet chestnut, 44
Sycamore wood as fuel, 46

T

Tables, firewood, 71–5
Table, percentage of water in timbers, 38
   , useful, 71
Timbers for charcoal gunpowder, 65–8
Timber of Acacia as fuel, 43
   Ailanthus as fuel, 43
   Alder as fuel, 43
   Alder buckthorn as fuel, 43
   Almond as fuel, 43
   Apple as fuel, 45
   Ash as fuel, 43
   Aspen as fuel, 38
   Beech as fuel, 44
   Birch as fuel, 44
   Black pine as fuel, 38
   Cedar of Lebanon as fuel, 44
   Cembra pine as fuel, 38
   Cherry as fuel, 44
   Chestnut, horse, as fuel, 44
   Chestnut, sweet, as fuel, 44
   Douglas fir as fuel, 40
   English elm as fuel, 38
   Hawthorn as fuel, 44
   Hazel as fuel, 45
   Hornbeam as fuel, 45
   Laburnum as fuel, 36
   Larch as fuel, 45
   Laurel as fuel, 37

Timber of Lime as fuel, 45
   Maple as fuel, 45
   Mountain pine as fuel, 38
   Mulberry as fuel, 45
   Oak as fuel, 45
   Pear as fuel, 45
   Pine as fuel, 45
   Plane as fuel, 45
   Poplar as fuel, 46
   Robinia as fuel, 38
   Scotch elm as fuel, 38
   Scotch pine as fuel, 38
   Scots fir as fuel, 37
   Silver fir as fuel, 38
   Spruce as fuel, 46
   Sycamore as fuel, 46
   Turkey oak as fuel, 38
   Walnut as fuel, 46
   Weymouth pine as fuel, 38
   Wild cherry as fuel, 37
   Willow as fuel, 46
   Yew as fuel, 46
Timbers, heating properties of uncommon, 40
   possessing great heating power, 38
   considerable heating power, 38
   fair heating power, 38
   little heating power, 38
Treatment of firewood, short notes on, 73
Tree stumps, 24
   roots as firewood, 21
   , blasting, 23
Trees, dead and dying, 20

U

Uncommon timbers, heating properties of, 40
Useful tables, 71
Utilisation of forest produce, 11

INDEX

V

firewoods, 35
oods, 43
oal, 60
specific gravity of, 72

W

fuel, 46
of, in timber, 38
d, 71
s fuel, 38

Wood for charcoal, comparative value of, 60
, home-grown, value of, 43
, nearest in heating properties to coal, 36
, scented, 41
Woods, best for fuel, 36
, various, heating properties of, 39–46

Y

Yew for firewood, 46

PRINTED IN GREAT BRITAIN BY
RICHARD CLAY & SONS, LIMITED,
BRUNSWICK ST., STAMFORD ST., S.E. 1,

University of California
SOUTHERN REGIONAL LIBRARY FACILITY
405 Hilgard Avenue, Los Angeles, CA 90024-1388
Return this material to the library
from which it was borrowed.

CPSIA information can be obtained
at www.ICGtesting.com
Printed in the USA
BVOW06s1927041216
469753BV00007B/88/P